General Aviation Pilot's Guide
Preflight Planning, Weather Self-Briefings, and Weather Decision Making

Foreword…………………………………………………………	ii
Introduction………………………………………………………	1
I Preflight Weather Planning…………………………………	**2**
Perceive – Understanding Weather Information………………	2
Process – Analyzing Weather Information ………………..……	7
Perform – Making A Weather Plan……………………………	11
II In-flight Weather Decision-Making………………………	**14**
Perceive – In-flight Weather Information………………............	14
Process – (Honestly) Evaluating In-flight Conditions…………..	16
Perform – Putting It All Together……………………………..	20
III Post-Flight Weather Review……………………………..	**22**
IV Resources……………………………………………………	**23**
Appendix 1 – Weather Products & Providers Chart……………..	24
Appendix 2 – Items for Standard Briefing………………………	25
Appendix 3 – Automated Weather Systems (definitions)…………	26
Appendix 4 – Developing Personal Weather Minimums. …………	27
Appendix 5 – Aviation Weather Analysis Worksheets………………	31
Appendix 6 – Weather Analysis Checklists (VFR)………………	32
Appendix 7 – Weather Analysis Checklists (IFR)………………	34
Appendix 8 – Estimating In-flight Visibility & Cloud Clearance……..	36

Foreword

This guide is intended to help general aviation (GA) pilots, especially those with relatively little weather-flying experience, develop skills in obtaining appropriate weather information, interpreting the data in the context of a specific flight, and applying the information and analysis to make safe weather flying decisions.

It has been developed with assistance and contributions from a number of weather experts, aviation researchers, air traffic controllers, and general aviation instructors and pilots. Special thanks are due to Dr. Dennis Beringer and Dr. William Knecht of the FAA's Civil Aviation Medical Institute (CAMI); Dr. Michael Crognale, Department of Psychology and Biomedical Engineering, University of Nevada/Reno; Dr. Douglas Wiegmann, Institute of Aviation, University of Illinois; Dr. B.L. Beard and Colleen Geven of the NASA Ames Research Center; Dr. Paul Craig, Middle Tennessee State University; Paul Fiduccia, Small Aircraft Manufacturers Association; Max Trescott, SJFlight; Arlynn McMahon, Aero-Tech Inc.; Roger Sharp, Cessna Pilot Centers; Anthony Werner and Jim Mowery, Jeppesen-Sanderson; Howard Stoodley, Manassas Aviation Center; Dan Hoefert; Lawrence Cole, Human Factors Research and Engineering Scientific and Technical Advisor, FAA; Ron Galbraith, FAA Air Traffic Controller, Denver ARTCC; Michael Lenz, FAA General Aviation Certification and Operations Branch, Christine Soucy, FAA Office of Accident Investigation; Dr. Rich Adams, Engineering Psychologist, FAA Flight Standard Service; and Dr. William K. Krebs, Human Factors Research and Engineering Scientific and Technical Advisor, FAA.

This guide is intended to be a living document that incorporates comments, suggestions, and ideas for best practices from GA pilots and instructors like you. Please direct comments and ideas to: susan.parson@faa.gov.

Happy – and safe – flying!

Introduction

Aviation has come a long way since the Wright brothers first flew at Kitty Hawk. One thing that has unfortunately not changed as much is the role that weather plays in fatal airplane accidents. Even after a century of flight, weather is still the factor most likely to result in accidents with fatalities.

From the safe perspective of the pilot's lounge, it is easy to second-guess an accident pilot's decisions. Many pilots have had the experience of hearing about a weather-related accident and thinking themselves immune from a similar experience, because "I would never have tried to fly in those conditions." Interviews with pilots who narrowly escaped aviation weather accidents indicate that many of the unfortunate pilots thought the same thing -- that is, until they found themselves in weather conditions they did not expect and could not safely handle.

Given the broad availability of weather information, why do general aviation (GA) pilots continue to find themselves surprised and trapped by adverse weather conditions? Ironically, the very abundance of weather information might be part of the answer: with many weather providers and weather products, it can be very difficult for pilots to screen out non-essential data, focus on key facts, and then correctly evaluate the risk resulting from a given set of circumstances.

 This guide describes how to use the **P**erceive – **P**rocess – **P**erform risk management framework as a guide for your preflight weather planning and in-flight weather decision-making. The basic steps are:

--**Perceive** weather hazards that could adversely affect your flight.

--**Process** this information to determine whether the hazards create risk, which is the potential impact of a hazard that is not controlled or eliminated.

-- **Perform** by acting to eliminate the hazard or mitigate the risk.

Let's see how the 3-P model can help you make better weather decisions.

Preflight Weather Planning

Perceive – Understanding Weather Information

When you plan a trip in a general aviation (GA) airplane, you might find yourself telling friends and family that you are first going to "see" if weather conditions are suitable. In other words, your first major preflight task is to *perceive* the flight environment by collecting information about current and forecast conditions along the route you intend to take, and then using the information to develop a good mental picture of the situation you can expect to encounter during the flight.

Because there are many sources of weather information today, the first challenge is simply knowing where and how to look for the weather information you need.

For many GA pilots, the FAA Flight Service Station (FSS) remains the single most widely used source of comprehensive weather information. Like other weather providers, the FSS bundles, or "packages," weather products derived from  National Weather Service (NWS) data and other flight planning information into a convenient, user-friendly package that is intended to offer the pilot not only specific details, but also a big picture view of the flight environment. In this respect, you might think of the FSS as "one-stop shopping" for GA weather information.

Flight Service offers four basic briefing packages:

- Outlook (for flights more than six hours away),
- Standard (for most flights),
- Abbreviated (to update specific items after a standard briefing); and
- TIBS (telephone information briefing service), which provides recorded weather information.

The specific weather information packaged into a standard briefing includes a weather synopsis, sky conditions (clouds), and visibility and weather conditions at the departure, en route, and destination points. Also included are adverse conditions, altimeter settings, cloud tops, dew point, icing conditions, surface winds, winds aloft, temperature, thunderstorm activity, precipitation, precipitation intensity, visibility obscuration, pilot reports (PIREPs), AIRMETs, SIGMETs, Convective SIGMETS, and Notices to Airmen (NOTAMs), including any temporary flight restrictions (TFRs).

Although a Flight Service weather briefing is still the single most comprehensive source of weather data for GA flying, it can be difficult to absorb all the information conveyed in a telephone briefing. Pictures are priceless when it comes to displaying complex, dynamic information like cloud cover and precipitation. For this reason, you may find it helpful to begin the preflight planning process by looking at weather products from a range of providers. The goal of this self-briefing process is to develop an overall mental picture of current and forecast weather conditions, and to identify areas that require closer investigation with the help of an FSS briefer.

Here is one approach to conducting your initial self-briefing. Keep in mind a simple rule-of-thumb as you work through the weather data collection process: the more doubtful the weather, the more information you need to obtain.

Television/Internet Sources. For long-range weather planning, many pilots start with televised or online weather, such as The Weather Channel (TWC) on television or the Internet. TWC is not an FAA-approved source of weather information, but its television and Internet offerings provide both tactical and strategic summaries and forecasts (up to 10 per day). TWC provides compact, easy-to-use information that can be a useful supplement to approved sources. For example, one TWC Internet page includes a weather map with color-coding for Instrument Flight Rules (IFR) and Marginal Visual Flight Rules (MVFR) conditions at airports around the country (http://www.weather.com/maps/aviation.html). This and other TWC features can give you a very useful first snapshot of weather conditions you will need to evaluate more closely. The National Weather Service's Aviation Weather Center (http://aviationweather.gov/) is another useful source of initial weather information. A look at the AIRMET and SIGMET watch boxes can quickly give you an idea of areas of marginal or instrument weather.

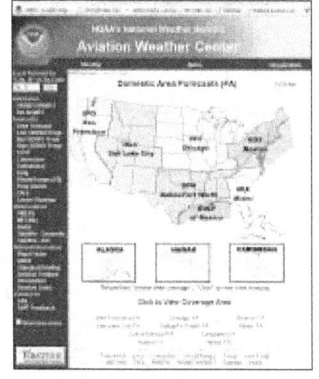

Direct User Access Terminal System (DUATS). Next, get a printed version of the FSS briefing package by obtaining a standard briefing for your route on DUATS. Free and accessible to all pilots via the Internet at www.duat.com (DTC) or www.duats.com (CSC), this resource provides weather information in an FAA-approved format and records the transaction as an official weather briefing. You might want to print out selected portions of the DUATS computer briefing for closer study and easy reference when you speak to a Flight Service briefer.

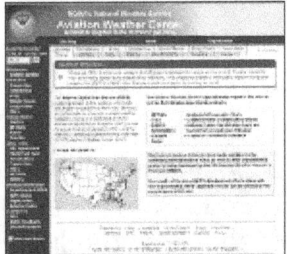

Aviation Digital Data Service (ADDS): You should also take a look at the wealth of weather information and resources available online via the Aviation Digital Data Service (ADDS), a joint effort of NOAA Forecast Systems Laboratory, NCAR Research Applications Program (RAP), and the National Centers for Environmental Prediction (NCEP) Aviation Weather Center (AWC). Available at http://adds.aviationweather.noaa.gov, ADDS combines information from National Weather Service (NWS) aviation observations and forecasts and makes them available on the Internet along with visualization tools to help pilots use this information for practical flight planning. For example:

- For METARs, TAFS, AIRMETS, and SIGMETS, the ADDS java tool can zoom in on specific parts of the country.

- For pilot reports (PIREPs), the ADDS Java tool can zoom in on a specific part of the country and specify the type of hazard reported (icing, turbulence, sky and weather). The tool also allows you to limit data to specified altitudes and time periods. Map overlays including counties, highways, VORs, and Air Route Traffic Control Boundaries are available.

- For the National Convective Weather Forecast (NCWF), the latest convection diagnostic is shown together with the one hour forecast. The java tool allows the user to select the height and speed of the forecasted thunderstorm, as well as the one-hour forecast from the previous hour to help the user understand how well the NCWF is performing.

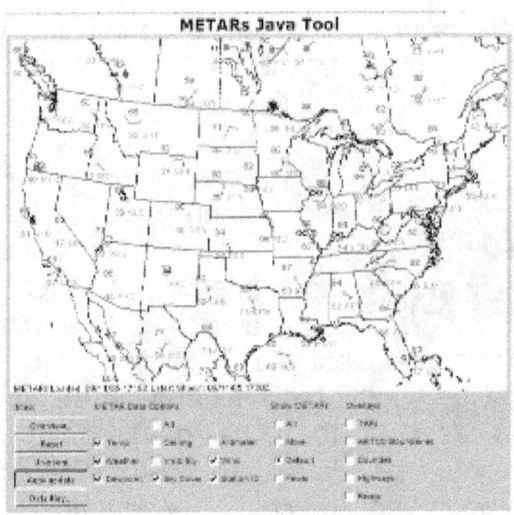

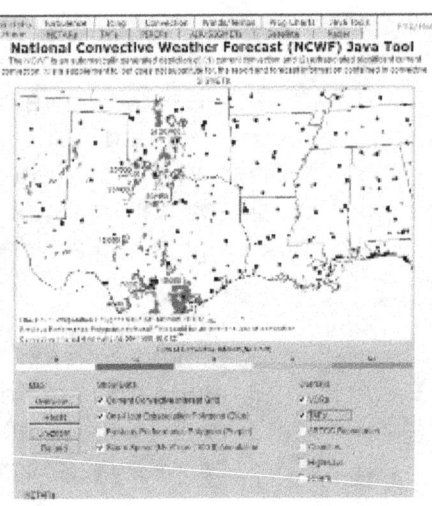

- ADDS also includes a Flight Path Tool that helps pilots visualize high resolution weather products together with winds aloft and pilot reports.

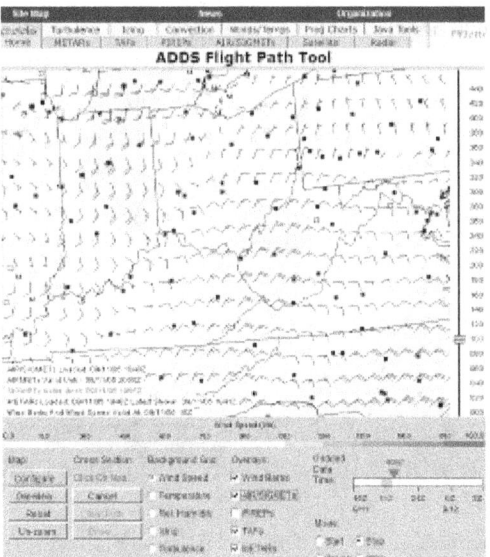

Although some of the other ADDS tools (e.g., icing potential and maximum turbulence potential) are only authorized for operational use by meteorologists and dispatchers, these products can still help you develop a mental picture of vertical and horizontal "weather hazard areas" for your flight.

Flight Service Station Briefing. Once you have formed a basic mental picture of the weather conditions for your trip, it is time to call the FSS. If you have just obtained a DUATS briefing or if the weather situation and mission are both simple, ask for an abbreviated briefing. If not, ask for a standard briefing. Armed with what you already know from your self-briefing process, you will find that it is much easier to absorb information from the briefer – and to know what questions you should ask.

A few guidelines for getting weather data from FSS:

✓ **DO** be sure to get the right FSS. When you dial the standard number, 1-800-WX-BRIEF from a cell phone, this number will connect you to the FSS associated with your cell phone's area code – not necessarily to the FSS nearest to your present position. If you are using a cell phone outside your normal calling area, check the *Airport/Facility Directory* to find the specific telephone number for the FSS you need to reach.

✓ **DO** know what you need, so you can request the right briefing "package" (outlook, standard, or abbreviated).

✓ **DO** use the standard flight plan form to provide the background the briefer needs to obtain the right information for you. Review the form before you call, and develop an estimate for items such as altitude, route, and estimated time en route so you can be sure of getting what you need to know.

Preflight Guide v. 1.3

- ✓ ***DO*** be honest – with yourself and with the briefer – about any limitations in pilot skill or aircraft capability.

- ✓ ***DO*** let the FSS specialist know if you are new to the area or unfamiliar with the typical weather patterns, including seasonal characteristics. If you are unfamiliar with the area, have a VFR or IFR navigation chart available while you listen to help sharpen your mental picture of where the weather hazards may be in relation to your departure airport, proposed route of flight, and destination.

- ✓ ***DO*** ask questions, and speak up if you don't understand something you have seen or heard. Less experienced pilots sometimes hesitate to be assertive. Smart pilots ask questions to resolve any ambiguities in the weather briefing. The worse the weather, the more data you need to develop options.

- ✓ ***DO*** be sure to get all the weather information you need. If you are flying in IMC or MVFR that could deteriorate, don't end the briefing without knowing which direction (north, south, east, west) to turn to fly toward better weather, and how far you would have to fly to reach it.

Process – Analyzing Weather Information

Obtaining weather information is only the first step. The critical next step is to study and evaluate the information to understand what it means for your circumstances.

The knowledge tests for most pilot certificates include questions on weather theory and use of weather products in aviation. However, it takes continuous study and experience to develop your skill in evaluating and applying weather data to a specific flight in a GA airplane. You might find it helpful to approach the task of practical, real world weather analysis with several basic concepts in mind.

What creates weather? Most pilots can recite the textbook answer -- "uneven heating of the earth's surface" – but what does that mean when you are trying to evaluate weather conditions for your trip? Let's take a look.

The three basic elements of weather are:

- *Temperature* (warm or cold);
- *Wind* (a vector with speed and direction); and
- *Moisture* (or humidity).

Temperature differences (e.g., uneven heating) support the development of low pressure systems, which can affect wide areas. Surface low pressure systems usually have fronts associated with them, with a "front" being the zone between two air masses that contain different combinations of the three basic elements (temperature, wind, and moisture).

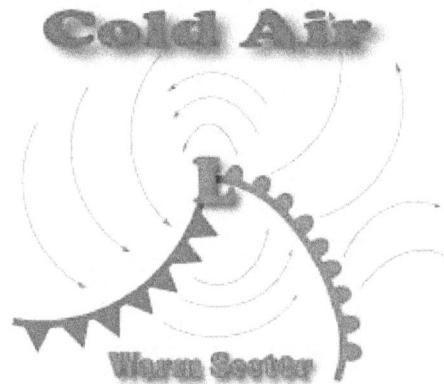

The illustration shows the "classic" northern hemisphere low pressure system with the associated cold and warm fronts. Remembering that air circulates counterclockwise around a low pressure system in the Northern Hemisphere will help you visualize the overall temperature, wind, and moisture patterns in a given area. Because weather is associated with fronts, which are in turn associated with low pressure systems, you can get some idea of possible conditions just by looking to see where the low pressure systems are in relation to your route.

What can weather do to you? Temperature, wind, and moisture combine to varying degrees to create conditions that affect pilots. The range of possible combinations is nearly infinite, but weather really affects pilots in just three ways. Specifically, the three basic weather elements can:

- *Reduce visibility*
- *Create turbulence*
- *Reduce aircraft performance*

How do you evaluate weather data? One approach to practical weather analysis is to review weather data in terms of how current and forecast conditions will affect visibility, turbulence, and aircraft performance for your specific flight.

Here's how it works. Suppose you want to make a flight from Cincinnati Municipal Airport (KLUK) to Port Columbus Airport in Columbus, Ohio (KCMH). You want to depart KLUK around 1830Z and fly VFR at 5,500 MSL. Your estimated time en route (ETE) is approximately one hour. Your weather briefing includes the following information:

METARs:

KLUK 261410Z 07003KT 3SM -RA BR OVC015 21/20 A3001
KDAY 261423Z 14005KT 3SM HZ BKN050 22/19 A3003
KCMH 261351Z 19005KT 3SM HZ FEW080 BKN100 OVC130 22/17 A3002

TAFs

KLUK 261405Z 261412 00000KT 3SM BR BKN015
   TEMPO 1416 2SM -SHRA BR
   FM1600 14004KT 5SM BR OVC035
   TEMPO 1618 2SM -SHRA BR BKN015
   FM1800 16004KT P6SM BKN040
   FM0200 00000KT 5SM BR BKN025
   TEMPO 0912 2SM BR BKN018

KDAY 261303Z 261312 06003KT 5SM BR SCT050 OVC100
   TEMPO 1315 2SM -RA BR BKN050
   FM1500 15006KT P6SM BKN050
   TEMPO 1519 4SM -SHRA BR BKN025
   FM1900 16007KT P6SM BKN035
   FM0200 14005KT 5SM BR BKN035
   FM0600 14004KT 2SM BR BKN012

KCMH 261406Z 261412 19004KT 4SM HZ SCT050 BKN120
   FM1800 17006KT P6SM BKN040
   TEMPO 1922 4SM -SHRA BR
   FM0200 15005KT 5SM BR BKN035
   FM0700 14004KT 2SM BR BKN012

WINDS ALOFT

	3000	6000	9000	12000	15000	18000	21000	24000	27000
CMH	1910	2108+15	2807+10	2712+05	2922-07	2936-17	294532	294540	313851
CVG	2310	2607+16	2811+11	2716+06	3019-05	2929-16	293430	293240	293652

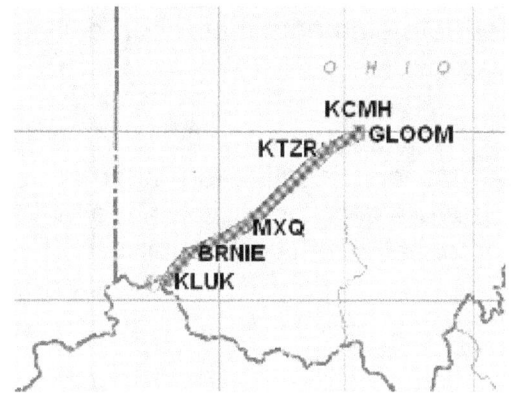

Remember that you have the option of getting this information in "plain English" format if you prefer not to decode. Whichever format you select, the first step is to look at your weather data in terms of three specific ways that weather can affect your flight: turbulence, visibility, and aircraft performance.

Organize the information into tables such as the one below, (*see Appendix 5 for blank forms*). This kind of format allows you to see and make "apples-to-apples" comparisons more easily. The column headings in the top row – arranged to match the order in which the briefing information is presented – can help you quickly identify the specific weather hazard(s) you might face on this trip. You may also find it helpful to convert Zulu (UTC) times to local time, and to write note expected ETAs for each waypoint on your flight plan.

Using the Cincinnati (KLUK) to Columbus (KCMH) trip as example:

CURRENT CONDITIONS

		Turbulence	Ceiling & Visibility			Visibility & Performance	Trends
Place	Time	Wind	Visibility	Weather	Ceiling	Temp/Dewpt	Altimeter
KLUK	1410Z	07003KT	3SM	RA, BR	OVC015	21/20	A3001
KDAY	1432Z	14005KT	3SM	HZ	BKN050	22/19	A3003
KCMH	1351Z	19005KT	3SM	HZ	FEW080, OVC130	22/17	A3002

FORECAST CONDITIONS

		Turbulence	Ceiling & Visibility		
Place	Time	Wind	Visibility	Weather	Ceiling
KLUK	FM1800Z	16004KT	P6 SM		BKN040
KDAY	TEMPO 1519Z	--	4SM	-SHRA	BKN025
	FM1900Z	16007KT	P6 SM	--	BKN035
KCMH	FM1800Z	17006KT	P6 SM	--	BKN040
	TEMPO 1922Z	--	4SM	-SHRA, BR	--

WINDS ALOFT

		Turbulence	Visibility & Performance
Place	Altitude	Wind	Temp
CVG	6000	260/07	16 C
CMH	6000	210/08	15 C

1. *Ceiling & Visibility.* First, look at the weather data elements that report ceiling and visibility.

In the case of the proposed VFR flight from KLUK to KCMH, current visibility at your departure and destination airports is marginal, and the small temperature/dew point spread should trigger a mental red flag for potentially reduced visibility. The forecasts call for conditions to improve at your departure airport, KLUK, by the time you plan to launch (1830Z).

Note, however, that you could encounter marginal conditions, including light rain showers, en route and also at your destination (KCMH). Since the forecast ceilings will probably not allow you to fly VFR at the planned altitude (5,500 MSL), this part of the analysis tells you that terrain and obstacle avoidance planning (discussed in the next section) will be necessary for this flight if you choose to depart at the originally scheduled time.

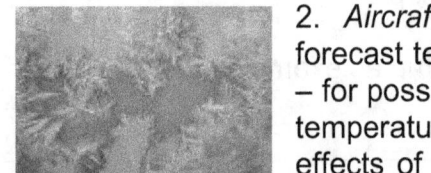

2. *Aircraft Performance.* Next, carefully review current and forecast temperatures – departure, en route, and destination – for possible adverse impact on aircraft performance. If the temperatures are high, you need to know and plan for the effects of high density altitude, especially on takeoff, climb, and landing. If temperatures are low and you plan on flying in the clouds, you should pay special attention to known or forecast icing and freezing levels.

In the sample VFR flight from KLUK to KCMH, temperatures on the surface and at your planned altitude are moderate, so performance problems associated with density altitude or icing are not likely to occur on this flight.

3. *Turbulence*: Review wind conditions for departure airport, en route, and destination airport. You will also need a mental picture of vertical wind profiles, so as to select the best altitude(s) for cruise flight, and to determine whether wind shear is present.

For the sample flight from KLUK to KCMH, the chart format allows you to see quickly that you will encounter light southerly surface winds at your departure and destination airports. Winds aloft will also be light, but from a westerly direction. There are no indications for wind shear or convective activity (thunderstorms), so you can conclude that turbulence is not likely to be a hazard for this particular flight.

For checklist questions and weather analysis worksheets to help you analyze the impact of these weather elements on your specific flight, see Appendix 6 (VFR) and Appendix 7 (IFR).

Perform – Making a Weather Plan

The third step in practical preflight weather planning is to perform an honest evaluation of whether your skill and/or aircraft capability are up to the challenge posed by this particular set of weather conditions. It is very important to consider whether the combined "pilot-aircraft team" is sufficient. For example, you may be a very experienced, proficient, and current pilot, but your weather flying ability is still limited if you are flying a 1980s-model aircraft with no weather avoidance gear. On the other hand, you may have a new technically advanced aircraft with moving map GPS, weather datalink, and autopilot – but if you do not have much weather flying experience, you must not count on the airplane's capability to fully compensate for your own lack of experience. You must also ensure that you are fully proficient in the use of onboard equipment, and that it is functioning properly.

One way to "self-check" your decision (regardless of your experience) is to ask yourself if the flight has any chance of appearing in the next day's newspaper. If the result of the evaluation process leaves you in any doubt, then you need to develop safe alternatives.

Think of the preflight weather plan as a strategic, "big picture" exercise. The goal is to ensure that you have identified all the weather-related hazards for this particular flight, and planned for ways to eliminate or mitigate each one. To this end, there are several items you should include in the weather flying plan:

Escape Options: Know where you can find good weather within your aircraft's range and endurance capability. Where is it? Which direction do you turn to get there? How long will it take to get there? When the weather is IMC (ceiling 1,000 or less and visibility 3 nm or less), identify an acceptable alternative airport for each 25-30 nm segment of your route. The worksheets in Appendices 5, 6, and 7 include space to record some of this information.

Reserve Fuel: Knowing where to find VFR weather does you no good unless you have enough fuel to reach it. Flight planning for only a legal fuel reserve could significantly limit your options if the weather deteriorates. More fuel means access to more alternatives. Having plenty of fuel also spares you the worry (and distraction) of fearing fuel exhaustion when weather has already increased your cockpit workload.

Terrain Avoidance: Know how low you can go without encountering terrain and/or obstacles. Consider a terrain avoidance plan for any flight that involves:

- Weather at or below MVFR (ceiling 1,000 to 3,000; visibility 3 to 5 miles)
- A temperature/dew point spread of 4° C. or less;
- Any expected precipitation; or
- Operating at night.

Know the minimum safe altitude for each segment of your flight. All VFR sectional charts include a maximum elevation figure (MEF) in each quadrangle. The MEF is determined by locating the highest obstacle (natural or man-made) in each quadrangle, and rounding up by 100 to 300 feet.

Charts for IFR navigation include a Minimum En route Altitude (MEA) and a 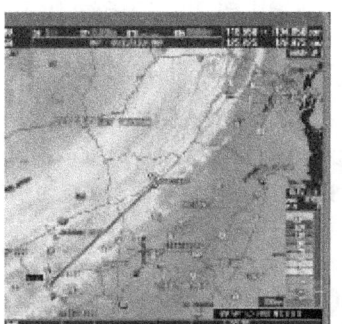 Minimum Obstruction Clearance Altitude (MOCA). Jeppesen charts depict a Minimum Off Route Altitude (MORA), while FAA/NACO charts show an Off Route Obstruction Clearance Altitude (OROCA) that guarantees a 1,000-foot obstacle clearance in non-mountainous terrain and a 2,000 foot obstacle clearance in mountainous terrain.

In addition to these sources, many GPS navigators (both panel-mount and handheld) include a feature showing the Minimum Safe Altitude (MSA), En route Safe Altitude (ESA), or Minimum En route Altitude (MEA) relative to the aircraft's position. If you have access to such equipment, be sure you understand how to access and interpret the information about safe altitudes.

The Air Safety Foundation's Terrain Avoidance Plan is another helpful resource.

Passenger Plan: A number of GA weather accidents have been associated with external or social pressures, such as the pilot's reluctance to appear "cowardly" or to disappoint passengers eager to make or continue a trip. There is almost always pressure to launch, and pressure to continue. Even the small investment in making the trip to the airport can create pressure to avoid "wasted" time.

 For this reason, your weather planning should include preflighting your passengers (and anyone waiting at your destination) as well as your aircraft. If you jointly plan for weather contingencies and brief your passengers before you board the aircraft, you as the pilot will be less vulnerable later on to the pressure to continue in deteriorating weather conditions. Suggestions:

✓ ***DO*** use the worksheet in Appendix 4 to develop personal minimums that will help you make the toughest go / no-go and continue / divert decisions well in advance of any specific flight.

✓ ***DO*** be aware that the presence of others can influence your decision-making and your willingness to take risks, and let your passengers know up front that safety is your top priority. Share your personal minimums with your passengers and anyone who might be waiting for you at the destination.

- ✓ **DO** establish "weather check" checkpoints every 25-30 nm along the route, at which you will reevaluate conditions. If possible, have your passengers assist by tracking progress and conditions at each weather checkpoint.

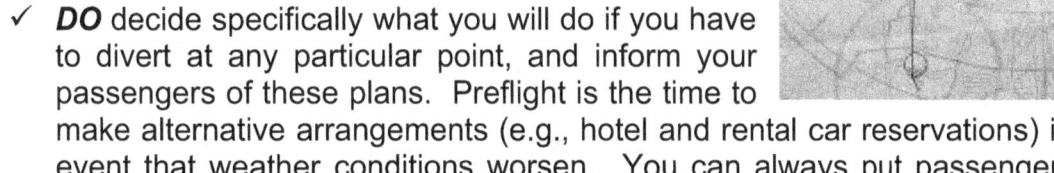

- ✓ **DO** use your pre-established personal minimums to determine exactly what conditions will trigger a diversion at any given weather checkpoint. Let your passengers know what these conditions are.

- ✓ **DO** decide specifically what you will do if you have to divert at any particular point, and inform your passengers of these plans. Preflight is the time to make alternative arrangements (e.g., hotel and rental car reservations) in the event that weather conditions worsen. You can always put passengers (or yourself) on an airliner if you absolutely have to return on time.

- ✓ **DO** advise anyone meeting you at your destination that your plans are flexible and that you will call them when you arrive. Be sure that they too understand that safety is your top priority, and that you will delay or divert if weather becomes a problem.

- ✓ **DO** remember that one of the most effective safety tools at your disposal is waiting out bad weather. Bad weather (especially involving weather fronts) normally does not last long, and waiting just a day can often make the difference between a flight with high weather risk and a flight that you can make safely.

In-flight Decision-Making

Perceive – Obtaining In-flight Weather Information

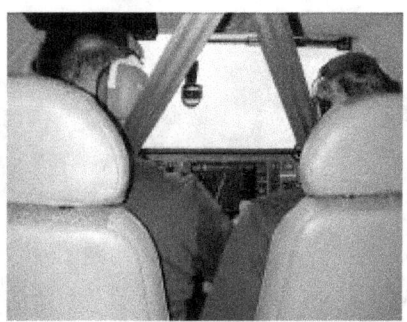

Many times, weather is not forecast to be severe enough to cancel the trip, so pilots often choose to take off and evaluate the weather as they go. While it is not necessarily a bad idea to take off and take a look, staying safe requires staying alert to weather changes. GA pilots and their aircraft operate in (rather than above) most weather. At typical GA aircraft speeds, making a 200-mile trip can leave a two to three hour weather information gap between the preflight briefing and the actual flight. In-flight updates are vital!

Let's take a closer look at in-flight weather data sources.

Visual Updates. One of the most important things you can do is to look outside. Use your eyes to survey the weather and literally see whether the conditions around you match the conditions that were reported or forecast. Sometimes there are local deviations in weather conditions (isolated cells, fog, etc.) that may not be immediately known to the FSS specialist or that may not appear on weather-product depictions, especially if there is no weather-reporting capability at your departure point. Even if you looked at radar during your preflight briefing process, remember that NEXRAD  data is at least 8 minutes old by the time you see it on a display, and older still by the time you are ready to depart. Weather can change very rapidly.

ATIS/ASOS/AWOS. One of the easiest ways to monitor conditions en route is to listen to ATIS and ASOS/AWOS broadcasts along your route. These broadcasts can help you update and validate preflight weather information about conditions along your route of flight.

En route Flight Advisory Service (EFAS, or Flight Watch). Available on 122.0 in the continental United States from 5,000 AGL to 17,500 MSL, EFAS, addressed as Flight Watch, is a service specifically designed to provide en route aircraft with timely and meaningful weather advisories pertinent to the type of flight intended, route of flight, and altitude. If you are in contact with ATC, request permission to leave the frequency to contact EFAS. Provide your aircraft identification and the name of the VOR nearest to your position.

Air Traffic Control (ATC). Simply monitoring ATC frequencies (available on aeronautical charts) along the way is one way to keep abreast of changing weather conditions. For example, are other GA aircraft along your route requesting diversions? You can also request information on the present location of weather, which the controller will try to provide if workload permits. When you ask ATC for weather information, though, you need to be aware that radar – the controller's primary tool – has limitations, and that operational considerations (e.g., use of settings that reduce the magnitude of precipitation returns) will affect what the controller can see on radar.

Datalink and Weather Avoidance Equipment. Radar and lightning detectors have been available in some GA aircraft for many years. These devices can contribute significantly to weather awareness in the cockpit. An increasing number of GA aircraft are now being equipped with weather datalink equipment, which uses satellites to transmit weather data such as METARs, TAFs, and NEXRAD radar to the cockpit, where it is often shown as an overlay on the multifunction display (MFD). Handheld devices with weather datalink capability are also a popular source of en route weather information.

There are several basic methods for transferring weather data from a weather data network provider to an aircraft:

- **Request/Reply** - In these systems, the pilot must decide what is needed and then request the specific information and coverage area. This request must then be sent from the aircraft to the satellite, from the satellite to the ground, processed by the ground system and transmitted back to the airplane. Transmission time can require as long as 10 or 15 minutes. Since weather can change very rapidly, this delay can significantly reduce utility of the data.

- **Narrowcast** - Some providers offer "narrowcast," which automatically sends data directly to the aircraft according to the pilot's pre-established preferences for products, update rate, resolution, coverage area, and other parameters.

- **Broadcast** - Broadcast systems continuously send available weather products to every user in the area through a satellite network and a system of interconnected ground stations. Satellite broadcast systems use high-power geosynchronous satellites to deliver large amounts of data in a very short time.

One of the most important, and critical, things to know about datalink is that regardless of the transmission method, it does not provide "real-time" information.

Process – (Honestly) Evaluating and Updating In-flight Conditions

Safe weather flying requires continuous evaluation of in-flight weather conditions.

Visual Updates. Seeing is believing – or so we are conditioned to think. Although you should certainly use your eyes during the flight to perceive the weather, you need to be aware that our prior visual experience largely determines our ability to "see" things. In the narrow runway illusion, for instance, the aircraft appears to be at a greater height over the runway because we have learned through previous experience what a typical runway should look like at a given altitude. The human brain prefers to adjust the apparent height of the aircraft rather than adjust the concept of what a runway should look like.

Similarly, scientists who study human vision have determined that weather transitions are sometimes too subtle for the limits of the visual system. Like other sensory organs, the eye responds best to changes. It adapts to circumstances that do not change, or those that change in a gradual or subtle way, by reducing its response. Just as the skin becomes so acclimated to the "feel" of clothing that it is generally not even noticed, the eye can become so accustomed to progressive small changes in light, color, and motion that it no longer "sees" an accurate picture. In deteriorating weather conditions, the reduction in visibility and contrast occurs quite gradually, and it may be quite some time before the pilot senses that the weather conditions have deteriorated significantly. In essence, you have to learn how to look past the visual illusion and see what is really there.

Certain weather conditions also make it particularly difficult to accurately perceive with the eye. For instance, a phenomenon called "flat light" can create very hazardous operating circumstances. Flat light is a condition in which all available light is highly diffused, and information normally available from directional light

sources is lost. The result is that there are no shadows, which means that the eye can no longer judge distance, depth features, or textures on the surface with any precision. Flat light is especially dangerous because it can occur with high reported visibility. It is common in areas below an overcast, and on reflective surfaces such as snow or water. It can also occur when blowing snow or sand create flat light conditions accompanied by "white-out," which is reduced visibility in all directions due to small particles of snow, ice or sand that diffuse the light.

Awareness is important in overcoming these challenges, but you can also develop your visual interpretation skills. Appendix 8 provides tips and techniques you can use to estimate in-flight visibility and cloud clearance, thus enhancing your ability to evaluate in-flight weather conditions accurately.

ATIS/ASOS/AWOS. In-flight weather information obtained from ATIS and ASOS/AWOS broadcasts can contribute useful pieces to the en route weather picture, but it is important to understand that this information is only a weather "snapshot" of a limited area. ATIS and ASOS/AWOS broadcasts are primarily intended to provide information on conditions in the airport vicinity. The information reported is derived from an array of sensors. While these systems are designed to be as accurate as possible and are increasingly sophisticated, the automated system is actually monitoring only a very small area on the airfield and that it reports only what it can "see." For example, sensors that measure visibility are actually measuring a section of air less than 24 inches wide. 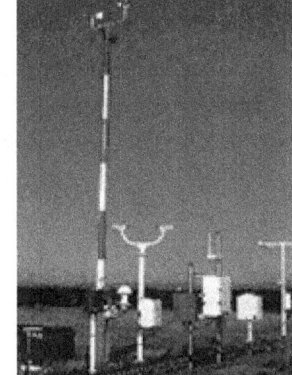 Even a dense fog on a portion of the airfield will go undetected by the system unless the fog actually obscures the sensors. The system will not "see" an approaching thunderstorm until it is almost directly over the automated site's ceiling instruments.

EFAS. Assuming that you do find or suspect deteriorating conditions while en route, be sure to contact the En route Flight Advisory Service (EFAS – Flight Watch) for additional information. EFAS can be an immensely helpful resource, but interpreting and applying the information you receive while you are also flying the aircraft – especially if you are in adverse or deteriorating conditions with no autopilot – can be very challenging. The key is understanding where the weather is in relation to your position and flight path, where it is going, and how fast it is moving. A good practice is to have an aeronautical chart with your route clearly marked readily available before you call Flight Watch. The chart will help you visualize where the weather conditions are in relation to your current position and intended route of flight, and determine whether (and where) you need to deviate from the original plan.

Another interpretation useful tool is the In-flight Advisory Plotting Chart (figure 7-1-2 in Chapter 7 of the Aeronautical Information Manual (AIM)). This chart includes the location and identifier for VORs and other locations used to describe hazardous weather areas. Consider keeping copies of this chart in your flight bag for easy reference whenever you call EFAS.

ATC. ATC radar can detect areas of precipitation, but does not detect clouds or turbulence. The existence of turbulence may be implied by the intensity of a precipitation return: the stronger the return, the more likely the presence of turbulence. Similarly, icing may be inferred by the presence of moisture, clouds, and precipitation at temperatures at or below freezing.

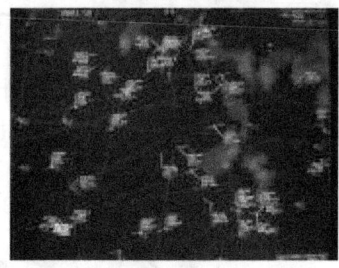

ARTCC facilities and many of the terminal approach control facilities now have digital radar display systems with processors that can better determine the intensity (dBZ) of radar weather echoes and display that information to the controller. Consequently, approach controllers, center controllers, and AFSS specialists have all begun using four terms to describe weather radar echoes to pilots: "light," "moderate," "heavy," and "extreme." Each term represents a precipitation intensity level paired with a decibel (dBZ) range to help pilots interpret the severity of the flight conditions present. (*Note: A dBZ is a measure of radar reflectivity in the form of a logarithmic power ratio with respect to radar reflectivity factor "Z."*)

Although the terms are consistent, there are still some equipment-related differences in what can be described.

- ✓ In Air Route Traffic Control Centers, NEXRAD data is fed through the **W**eather **a**nd **R**adar **P**rocessor (WARP), which organizes 16 NEXRAD levels into four reflectivity (dBZ) categories. Reflectivity returns of less than 30 dBZ are classified as "LIGHT" and are filtered out of the center controllers' display, which means that center controllers cannot report areas of "light" weather radar echoes.

- ✓ A terminal radar approach control has neither NEXRAD nor WARP, so weather radar echoes are displayed by the Airport Surveillance Radar (ASR) systems using Common Automated Radar Terminal System (Common ARTS) or Standard Terminal Automation Replacement System (STARS) digital weather processors. Paired with a weather processor, digitized ASR 9 and 11 systems display the four weather radar echo intensity categories to the controller.

- ✓ Terminal radar approach control facilities can, and do, display "light" (less than 30 dBZ) areas of precipitation. Not all terminal facilities have digitized systems, however, and systems without digital processors cannot discern radar echo intensity. In these cases, ATC can describe the position of weather radar echoes, but will state "intensity unknown" instead of using the terms, "light," "moderate," "heavy," or "extreme."

A critical element in interpreting weather information from ATC is a thorough understanding of pilot-controller communications. Be sure to review the AIM Pilot/Controller Glossary, and clarify points you do not understand.

Datalink and Weather Avoidance Equipment. When analyzing this information, it is vital to remember that the quality of the information depends heavily upon update rate, resolution, and coverage area. When flying an aircraft that has datalink equipment, safe and accurate interpretation of the information you receive depends on your understanding of each of these parameters.

Datalink does not provide real-time information. Although weather and other navigation displays can give pilots an unprecedented quantity of high quality weather data, their use is safe and appropriate only for *strategic* decision making (attempting to avoid the hazard altogether). **Datalink is not accurate enough or current enough to be safely used for *tactical* decision making** (negotiating a path through a weather hazard area, such as a broken line of thunderstorms).

Be aware that onboard weather equipment can inappropriately influence your decision to continue a flight. No matter how "thin" a line of storms appears to be, or how many "holes" you think you see on the display, it is not safe to fly through them.

Perform – Putting It All Together

In the preflight planning process, you used weather data and analysis to develop a strategic, "big picture" weather flying plan. During the en route phase, use the data and analysis to make tactical weather decisions. Good tactical weather flying requires you to perceive the conditions around you, process (interpret) their impact on your flight, and perform by taking appropriate action at each stage.

✓ **DO** reassess the weather on a continuous basis. Designate specific fixes (e.g., airports) on or near your flight path as "weather check" checkpoints and use one of the in-flight resources described above to get updated information.

✓ **DO** take action if you see or suspect deteriorating weather:

- Trust your eyes if you see weather conditions deteriorating.

- Contact EFAS for detailed information.

- Head for the nearest airport if you see clouds forming beneath your altitude, gray or black areas ahead, hard rain or moderate turbulence, or clouds forming above that require you to descend. It is much easier to reevaluate conditions and make a new plan from the safety of an airport.

✓ **DO** contribute to the system by making pilot reports (PIREPS) when you call Flight Watch. To learn more about making good PIREPS, take the Air Safety Foundation's free online "Skyspotter" course.

ATC. If you need help from ATC in avoiding or escaping weather, ask sooner rather than later. Guidelines:

✓ **DO** be sensitive to ATC communications workload, but keep controllers advised of your weather conditions. Tell the controller if you need to deviate.

✓ **DO** remember that navigational guidance information issued to a VFR flight is *advisory* in nature. Suggested headings do not authorize you to violate regulations, and they are not guaranteed to keep you clear of all weather.

✓ **DON'T** hesitate to ask questions if you do not understand or if you are unsure.

✓ **DON'T** make assumptions about what the controller knows about your flight:

- If you need ATC's help to avoid convective weather, it never hurts to remind the controller that you have no onboard weather avoidance equipment.

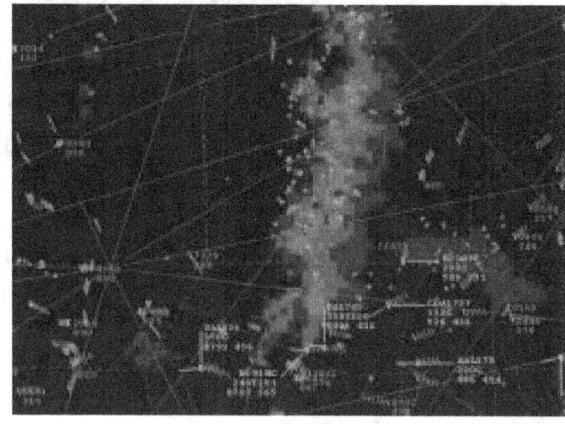

- If you are handed off while on a suggested heading for weather avoidance, confirm that the next controller knows you are requesting weather avoidance assistance. For example, your initial call might be: "Center, N2817S, level 5,000, zero two zero heading for weather avoidance."

- ***Never*** assume that "cleared direct when able" means that flying a direct course at that time will keep you clear of weather. To ATC, "direct when able" means to fly direct when you are able to receive a signal and navigate directly to the fix. If you have any doubt, ASK whether a direct course will keep you clear of areas with moderate and heavy radar returns indicative of thunderstorm activity.

- Words such as "showers" and "precipitation" can be very misleading. Some pilots mistakenly assume that these words indicate areas of rain with no thunderheads present. In the world of ATC, weather radar echoes are all referred to as "precipitation." Do not proceed into areas of "showers" or "precipitation" without clarifying whether the level of precipitation is "light," "moderate," or "heavy."

DON'T terminate VFR flight following or other services and leave an ATC frequency without informing the controller that you are doing so.

Post-Flight Weather Review

When you land after a challenging flight in the weather, you may want nothing more than to go home and unwind. The immediate post flight period, however, is one of the best opportunities to increase your weather knowledge and understanding. Studies show that pilots sometimes fly into bad weather simply because they lack relevant experience, and thus did not recognize that certain weather "cues" might create a safety hazard to the flight. Make it a point to learn something from every weather encounter. At the end of a flight involving weather, take a few minutes to mentally review the flight you just completed and reflect on what you learned from this experience. To guide your post flight weather review:

✓ What weather conditions/hazards existed, and how did they impact this flight?

    Turbulence / Winds _____
    Ceilings / Visibility _____
    Aircraft Performance _____

✓ How did the conditions encountered during this flight compare with the information obtained in the preflight briefing?

✓ Which source(s) of preflight weather information provided the best (or most useful, most accurate, most relevant) data for this flight?

✓ Which source(s) of en route weather information provided the best (or most useful, most accurate, most relevant) data for this flight?

Another way to develop your weather experience and judgment is simply to observe and analyze the weather every day. When you look out the window or go outside, observe the clouds. What are they doing? Why are they shaped as they are? Why is their altitude changing? This simple habit will help you develop the ability to "read" clouds, and understand how shape, color, thickness, and altitude can be valuable weather indicators. As your cloud-reading skill develops, start trying to correlate the temperature, dew point, humidity, and time of day to the types of clouds that have formed. Take note of the wind and try to visualize how it wraps around the tree or whips around the corner of a building. This exercise will help you become more aware of wind at critical points in your flight.

Weather is a fact of life for pilots. Developing your weather knowledge and expertise is well worth the time and effort you put into it, because weather wisdom will help keep you – and your passengers – safe in the skies.

Resources

Appendix 1   Weather Products and Weather Providers Chart

Appendix 2   Items for Standard Briefing

Appendix 3   Automated Weather Observing Systems

Appendix 4   Developing Personal Weather Minimums

Appendix 5   Aviation Weather Analysis Worksheets

Appendix 6   Weather Analysis Checklist - VFR

Appendix 7   Weather Analysis Checklist - IFR

Appendix 8   Estimating In-flight Visibility and Cloud Clearance

Appendix 1
Weather Products and Weather Providers

The table below lists some of the most common weather products and providers:

Source	AC (Severe Wx Outlook)	AIRMET / SIGMET	charts, Convective outlook	charts, Prog.	charts, Radar (NEXRAD)	charts, Radar summary	charts, Surface analysis	Center Weather Advisory (ATC)	charts, Weather depiction	FA (18-hr area forecast)	FD (winds/temps aloft forecast)	FD (winds/temps aloft forecast)	METAR	PIREP	Satellite	SD (hourly radar)	TAFs	TWEB
	Format: T = text; G = Graphic. Text may be written or spoken.																		
Preflight	T	T	G	G	G	G	G	T	G	T	T	G	T	T	G	T	T	T
Commercial vendor	Search Internet for "commercial weather products."
Public NWS or NOAA site		X	X	X		X			X		X	X	X	X	X	X	X	X
ADDS (aviation digital data)	X	X	X	X	X	X	X	X	X	X	X	X	X	X	X	X	X	
DUATS	X	X					X	X	X	X		X	X	X	X	X		
FSS (automated TIBS)	Short automated briefing, origin & radius, advisories & summary, ceil, vis, w. easy link to FS Specialist
FSS (standard)	Verbal synopsis of all available information
FSS (abbreviated)	Short, verbal synopsis, based on all available information
FSS (outlook)	Short, verbal forecast based on all available information
The Weather Channel					X								X		X			
En route																		
cockpit avionics	products vary
EFAS	Verbal synopsis, based on all available information
HIWAS		X						X							X			
TWEB	Short automated synopsis, origin & radius, wx advisories, ceil, vis, winds, radar, PIREPS, alerts
Both																		
ASOS	rowspan: ASOS, ATIS, AWOS are similar to METAR, incl. Place, Time, Wind direction/speed, Visibility, Ceiling, Temp/Dewpoint, Altimeter
ATIS	
AWOS	
CWA	Short, verbal synopsis, based on all available information

(NOTE: Products directly accessible to the user are marked with an "X.")

ADDS	Aviation Digital Data Service (ADDS) (http://adds.aviationweather.noaa.gov/)
ASOS	Automated Surface Observing System
ATIS	Automated Terminal Information Service
AWOS	Automated Weather Observing System
CWA	Center Weather Advisory
DUATS	Direct User Access Terminal System
EFAS	En route Flight Advisory System
FSS	Flight Service Station
HIWAS	Hazardous In-flight Weather Advisory System
LLWAS	Low Level Wind Shear Alert System
NOAA	National Oceanic and Atmospheric Association
NWS	National Weather Service
TIBS	Telephone Information Broadcast Service
TWEB	Transcribed Weather Broadcast

Appendix 2
Items for Standard Briefing

- ✓ Type of Flight (VFR or IFR)
- ✓ Aircraft identification
- ✓ Aircraft Type / Special Equipment
- ✓ True Airspeed
- ✓ Departure Point
- ✓ Proposed Departure Time
- ✓ Cruising Altitude
- ✓ Route of Flight
- ✓ Destination
- ✓ Estimated Time En Route
- ✓ Remarks (e.g., "no weather avoidance equipment on board")
- ✓ Fuel
- ✓ Alternate Airports
- ✓ Pilot's Name

Appendix 3
Automated Weather Observing Systems

AWOS- Automated Weather Observing System.

ASOS- Automated Surface Observing System.

AWOS-3 reports all the items in a METAR – time of observation, wind, visibility, sky coverage/ceiling, temperature, dew point and altimeter setting. The designator "**A02**" in the remarks portion of the observation indicates the station has a precipitation discriminator that determines the difference between liquid and freezing/frozen precipitation.

ASOS reports the same data as AWOS-3 **plus** precipitation type and intensity like the AWOS-3 sites with the A02 capabilities.

AWOS-2 reports the same METAR items as an AWOS-3 except it **does not report sky coverage/ceiling information.**

AWOS-1 reports the time of observation, wind, temperature, dew point and altimeter setting. **It does not report visibility or sky coverage information.**

AWOS-A reports **only** the time of observation and altimeter setting.

The prefix "AUTO" indicates the data is derived from an automated system. A certified weather observer may provide augmented weather and obstruction to visibility information in the remarks of the report at AWOS locations.

The "AUTO" prefix disappears when the report has been augmented by human observers.

Federal Aviation Administration

Developing *Personal* Minimums

Think of personal minimums as the human factors equivalent of reserve fuel. Personal minimums should provide a solid safety buffer between:

- *Skills required* for the specific flight, and

- *Skills available* to you through your training, experience, currency, and proficiency.

Step 1 – Review Weather Minimums

Step 2 – Assess Weather Experience and Personal Comfort Level

Step 3 – Consider Winds and Performance

Step 4 – Assemble Baseline Values

Step 5 – Adjust for Specific Conditions

Step 6 – Stick to the Plan!

Step 1: Review definitions for VFR & IFR weather minimums.

Category	Ceiling		Visibility
VFR	greater than 3,000 AGL	and	greater than 5 miles
MVFR	1,000 to 3,000 AGL	and/or	3 to 5 miles
IFR	500 to 999 AGL	and/or	1 mile to less than 3 miles
LIFR	below 500 AGL	and/or	less than 1 mile

Step 2(a): Record certification, training, & recent experience.

CERTIFICATION LEVEL	
Certificate level (e.g., private, commercial, ATP)	
Ratings (e.g., instrument, multiengine)	
Endorsements (e.g., complex, HP, high altitude)	
TRAINING SUMMARY	
Flight review (e.g., certificate, rating, Wings)	
Instrument Proficiency Check	
Time since checkout in airplane 1	
Time since checkout in airplane 2	
EXPERIENCE	
Total flying time	
Years of flying experience	
RECENT EXPERIENCE (last 12 months)	
Hours	
Hours in this airplane (or identical model)	
Normal Landings	
Crosswind landings	
Night hours	
Night landings	
Hours flown in high density altitude	
Hours flown in mountainous terrain	
IFR hours	
IMC hours (actual conditions)	
Approaches (actual or simulated)	
Time with specific GPS navigator	
Time with specific autopilot	

Step 2(b): Enter values for weather experience/ "comfort level."

Experience & "Comfort Level" Assessment Combined VFR & IFR					
Weather Condition		VFR	MVFR	IFR	LIFR
Ceiling					
	Day				
	Night				
Visibility					
	Day				
	Night				

Step 3(a): Enter values for experience / comfort in turbulence.

Experience & "Comfort Level" Assessment Wind & Turbulence			
	SE	ME	Make/ Model
Turbulence			
Surface wind speed			
Surface wind gusts			
Crosswind component			

Step 3(b): Enter values for performance.

Experience & "Comfort Level" Assessment Performance Factors			
	SE	ME	Make/ Model
Performance			
Shortest runway			
Highest terrain			
Highest density altitude			

Step 4: Assemble and evaluate baseline personal minimums.

Baseline Personal Minimums

Weather Condition		VFR	MVFR	IFR	LIFR
Ceiling					
	Day				
	Night				
Visibility					
	Day				
	Night				

Turbulence		SE	ME	Make/Model	
	Surface Wind Speed				
	Surface Wind Gust				
	Crosswind Component				

Performance		SE	ME	Make/Model	
	Shortest runway				
	Highest terrain				
	Highest density altitude				

Step 5: Adjust for specific conditions.

	If you are facing:		Adjust baseline personal minimums to:
Pilot	Illness, medication, stress, or fatigue; lack of currency (e.g., haven't flown for several weeks)	Add	At least 500 feet to ceiling
			At least ½ mile to visibility
Aircraft	An unfamiliar airplane, or an aircraft with unfamiliar avionics/equipment:		At least 500 ft to runway length
enVironment	Airports and airspace with different terrain or unfamiliar characteristics	Subtract	At least 5 knots from winds
External Pressures	"Must meet" deadlines, passenger pressures; etc.		

Appendix 5
Aviation Weather Analysis Forms

CURRENT CONDITIONS (from METARs)

Place	Time	Turbulence	Ceiling & Visibility			Visibility & Performance	Trends
		Wind	Visibility	Weather	Ceiling	Temp/Dewpt	Altimeter

FORECAST CONDITIONS (from TAFs)

Place	Time	Turbulence	Ceiling & Visibility		
		Wind	Visibility	Weather	Ceiling

WINDS ALOFT

Place	Altitude	Turbulence	Visibility & Performance
		Wind	Temp

Preflight Guide v. 1.3

Appendix 6
Weather Analysis Checklists – VFR Flight

Ceiling & Visibility

- ✓ How much airspace do I have between the reported/forecast ceilings and the terrain along my route of flight? Does this information suggest any need to change my planned altitude?

- ✓ If I have to fly lower to remain clear of clouds, will terrain be a factor?

- ✓ How much ground clearance will I have?
- ✓ Do I have reliable ceiling information?
- ✓ Will I be over mountainous terrain or near large bodies of water where the weather can change rapidly, or where there may not be a nearby weather reporting station?

- ✓ What visibility can I expect for each phase of flight (departure, enroute, destination)?

- ✓ Given the speed of the aircraft, expected light conditions, terrain, and ceilings, are the reported and forecast visibility conditions sufficient for this trip?

- ✓ Are there conditions that could reduce visibility during the planned flight? (Hint: look for indications such as a small and/or decreasing temperature/dew point spread).

- ✓ Are reported and forecast ceiling & visibility values above my personal minimums?

Aircraft Performance

- ✓ Given temperature, altitude, density altitude, and aircraft loading, what is the expected aircraft performance?

  o Takeoff distance
  o Time & distance to climb
  o Cruise performance
  o Landing distance

- ✓ Are these performance values sufficient for the runways to be used and the terrain to be crossed on this flight?

(*Remember that it is always good practice to add a 50% to 100% safety margin to the "book numbers" you derive from the charts in the aircraft's approved flight manual (AFM)*).

Turbulence

- ✓ Are the wind conditions at the departure and destination airports within the gust and crosswind capabilities of both the pilot and aircraft? (*Note: For most GA pilots, personal minimums in this category might be for a maximum gust of 5 knots and maximum crosswind component 5 knots below the maximum demonstrated crosswind component.*)

- ✓ What is the maneuvering speed (V_A) for this aircraft at the expected weight?

(*Note: Remember that V_A is lower if you are flying at less than maximum gross weight.*)

VFR Analysis Worksheet

Place	Time	Ceiling & Visibility			Visibility & Performance		Trends
		Wind	Visibility	Weather	Ceiling	Temp/Dewpt	Altimeter

Column group headers: Turbulence (Wind), Ceiling & Visibility (Visibility, Weather, Ceiling), Visibility & Performance (Temp/Dewpt), Trends (Altimeter).

Turbulence Analysis

Personal Minimums:
- Wind speed = _____
- Gust factor = _____
- Crosswind = _____

Departure wind = _____ @ _____
Destination wind = _____ @ _____
En route wind = _____ @ _____
Maneuvering speed = _____ *

Convective SIGMETS? Yes ☐ No ☐

Nearest Good Weather

Direction: N S E W

Distance: _____ nm

Flying time to nearest good VFR: _____

* V_A decreases as weight decreases

Ceiling and Visibility Analysis

Personal Minimums:
- Ceiling = _____
- Visibility = _____

Planned altitude = _____
- Lowest en route ceiling = _____   } ground clearance

Planned altitude = _____
- Highest en route obstacle = _____   } clearance

Planned altitude = _____
- Highest en route terrain = _____   } clearance

AIRMETS? Yes ☐ No ☐
SIGMETS? Yes ☐ No ☐
Reliable ceiling information? Yes ☐ No ☐
Over mountainous terrain? Yes ☐ No ☐
Over large bodies of water? Yes ☐ No ☐

Departure visibility = _____
Lowest en route visibility = _____
Destination visibility = _____

Performance Analysis

Density altitude = _____
Freezing level = _____

Takeoff distance = _____
Runway length = _____

Landing distance = _____
Runway length = _____

Cruise performance = _____

Fuel available = _____ gal _____ hrs
Fuel required = _____ gal _____ hrs
Fuel reserve = _____ gal _____ hrs

Note: It is good practice to add a 50% to 100% safety margin to the "book numbers" you derive from charts in the approved flight manual (AFM).

Appendix 7
Weather Analysis Checklist – IFR Flight

Ceiling and Visibility

- ✓ Is the forecast ceiling for my estimated time of arrival high enough to make the approach?

- ✓ What visibility can I expect for each phase of flight (departure, enroute, destination)?

    --Will I have enough visibility to legally make an instrument approach at the destination?

    --Do current or forecast ceiling and visibility conditions require me to select and file an alternate? (1-2-3 rule.)

    --Where is the nearest GOOD weather alternative?

- ✓ How do reported and forecast conditions for ceiling and visibility compare with my personal minimums for IFR?

Aircraft Performance

- ✓ Given temperature, altitude, density altitude, and aircraft loading, what is the expected aircraft performance?

    o Takeoff distance
    o Time & distance to climb
    o Cruise performance
    o Landing distance

- ✓ Are these performance values sufficient for the runways to be used and the terrain to be crossed on this flight?

    (Remember that it is always good practice to add a 50% to 100% safety margin to the "book numbers" you derive from the charts in the aircraft's approved flight manual (AFM)).

- ✓ Will weight restrictions allow me to carry more than the normal fuel reserve?

    (More fuel means that you have more options to escape weather.)

- ✓ *Icing.* What is the forecast freezing level for this flight?

    o Are there any pilot reports (PIREPS) for my route, or points on the route that support or rebut the icing forecast?

    o Where are the cloud bases and cloud tops?

Turbulence

- ✓ Are the wind conditions at the departure and destination airports within the gust and crosswind capabilities of both the pilot and aircraft?

- ✓ What is the maneuvering speed (V_A) for this aircraft at the expected weight?

    (Remember that V_A is lower if you are flying at less than maximum gross weight.)

- ✓ *Thunderstorms.* Does the forecast include convective activity at any point along my proposed route?

IFR Analysis Worksheet

Place	Time	Ceiling & Visibility			Turbulence	Visibility & Performance	Trends
		Visibility	Weather	Ceiling	Wind	Temp/Dewpt	Altimeter

Turbulence Analysis

Personal Minimums:
- Wind speed = _____
- Gust factor = _____
- Crosswind = _____

Departure wind = _____ @ _____
Destination wind = _____ @ _____
En route wind = _____ @ _____
Maneuvering speed = _____ *

T-storms forecast? Yes ☐ No ☐
Convective SIGMETS? Yes ☐ No ☐

Nearest VFR Weather

Direction: N S E W

Distance: _____ nm

Flying time to nearest good VFR: _____

* V_A decreases as weight decreases

Ceiling and Visibility Analysis

Personal IFR Approach Minimums:
- Ceiling = _____
- Visibility = _____

Planned altitude = _____
- Lowest en route ceiling = _____ } ground clearance

Planned altitude = _____
- Highest en route obstacle = _____ } clearance

Planned altitude = _____
- Highest en route terrain = _____ } clearance

Cloud bases = _____ Cloud tops = _____

Alternate required ? Yes ☐ No ☐
Over mountainous terrain ? Yes ☐ No ☐
Over large bodies of water ? Yes ☐ No ☐

Departure visibility = _____
Lowest en route visibility = _____
Destination visibility = _____

Performance Analysis

Density altitude = _____
Freezing level = _____

Takeoff distance = _____
Runway length = _____

Landing distance = _____
Runway length = _____

Cruise performance = _____

Fuel available = _____ gal _____ hrs
Fuel required = _____ gal _____ hrs
Fuel reserve = _____ gal _____ hrs

Note: It is good practice to add a 50% to 100% safety margin to the "book numbers" you derive from charts in the approved flight manual (AFM).

Appendix 8
Estimating In-flight Visibility & Cloud Clearance

There are a number of ways to develop your skill in estimating your in-flight visibility and cloud clearance. These techniques will help you establish a continuous weather assessment habit. It will also help you calibrate your perceptions and learn when to trust what you see.

- ✓ Listen to the ATIS or ASOS/AWOS as you pass near an airport. First try to evaluate the basic weather conditions based on what you see. Then listen to the ATIS or ASOS/AWOS and compare the official report to your own evaluation of conditions, as well as with any previous reports you have seen from this location.

- ✓ Use the length of a runway you pass in flight to estimate distances.

    - o A runway that is 5,300 feet long is about a mile. Look to see how far ahead you can see, and estimate the number of runways that it would take to cover that distance.

    - o A 2,600 foot runway would be about a half mile, and so on. In this case, visibility is less than 3 miles if you cannot see 6 runway lengths ahead.

- ✓ If you know your aircraft's groundspeed, you can estimate distance. Look to the most distant point you can see ahead and then time how long it takes to reach it.

    - o If, for example, your ground speed is 105 knots, that's about 120 mph and you'll cover about 2 miles per minute. If you reach the point in less than 90 seconds, the in-flight visibility is less than 3 miles!

    - o A simple variation on this technique it to use GPS or DME while flying directly to or from a waypoint or VOR. Just look at the beginning and ending mileage on the GPS or DME to see how far you've flown to reach the farthest point you can see.

- ✓ If you need to know the lateral distance to a cloud, start timing when the cloud is ahead of you and at about a 45° angle (halfway between your 10 and 11 o'clock or between your 1 and 2 o'clock positions). Stop timing when the cloud is off your wingtip. The distance you've traveled forward will now be equal to the distance between you and the cloud. If you were traveling at 120 mph, it will take you about 11 seconds to travel 2000 feet. If the cloud took less than 11 seconds to arrive off your wingtip, you are now less than 2000 feet horizontally from that cloud.

(courtesy of Max Trescott, SJ Flight)

www.ingramcontent.com/pod-product-compliance
Lightning Source LLC
Chambersburg PA
CBHW081805170526
45167CB00008B/3330